AMAZING BUT TRUE

By the editors of OWL Magazine

Greey de Pencier Books

©1987, Greey de Pencier books; originally published as OWL's Amazing But True #1, ©1983. No part of this book may be reproduced or copied in any form without written permission from the publisher. Second printing 1988; third printing 1989.

ISBN 0-919872-80-8

TM: OWL and the OWL character are trademarks of the Young Naturalist Foundation.

Books from OWL are published in Canada by Greey de Pencier Books, 56 The Esplanade, Suite 306, Toronto, M5E 1A7.

Canadian Cataloguing in Publication Data
Main entry under title:
Amazing but True

First published, 1983, under title:
Owl's amazing but True, v. 1.
ISBN 0-919872-80-8

1. Animals – Miscellanea – Juvenile literature.
2. Nature – Miscellanea – Juvenile literature. I. Dingwall, Laima, 1953- . II. Slaight, Annabel, 1940- . III. Title: Owl's amazing but true.

QL49.A62 1987   570   C87-093423-6

Printed in Canada.

Edited by Laima Dingwall
Cover design by Wycliffe Smith
Cover illustration by Vesna Krystanovich

Introduction

We at OWL Magazine have been gathering wonderful facts about the world for many years, reading books and talking to all sorts of people from kids to scientists. At last, we've put together some of our favorite facts so that you can be as amazed about the world as we are.

You probably know that sharks, whales, dragons and tigers are interesting. When you learn about them in this book you'll discover that, spiders, beetles, birds and even trees can be every bit as fascinating.

We made this *Amazing But True* book pocket-sized so you can pick it up, look at it and think about the things you discover whenever you have a little extra time. And, if you're in a real hurry for information, you can use the handy index in the back.

Step into Our Parlor

Most spiders you see sit around waiting for prey to get caught in their webs. But others are much more energetic hunters....

The wolf spider like its namesake, the wolf, runs after dinner and seldom gives up the chase until it has snapped its powerful jaws around its meal.

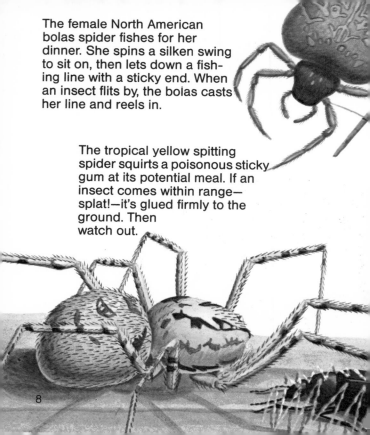

The female North American bolas spider fishes for her dinner. She spins a silken swing to sit on, then lets down a fishing line with a sticky end. When an insect flits by, the bolas casts her line and reels in.

The tropical yellow spitting spider squirts a poisonous sticky gum at its potential meal. If an insect comes within range—splat!—it's glued firmly to the ground. Then watch out.

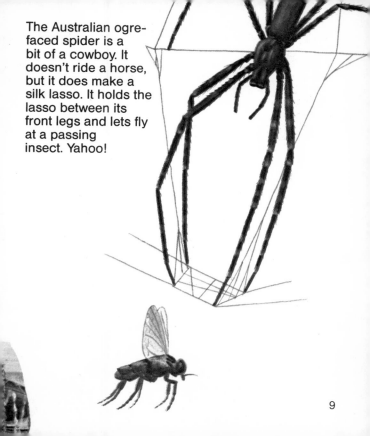

The Australian ogre-faced spider is a bit of a cowboy. It doesn't ride a horse, but it does make a silk lasso. It holds the lasso between its front legs and lets fly at a passing insect. Yahoo!

Wonderful Webs

Spider webs are sticky because the spiders that spin them use them to trap prey. How do web-weaving spiders walk on the webs without getting stuck? They have oily feet that slip and slide easily over the silk. Should a spider fall backward onto its web, it will stick there like any other creature. Aargh!

Big Web

If you are ever in India, you'll know when you've seen the Indian orbweaver spider's home. Its web is so big that it could cover half your classroom.

Small Web

The Glyphesis cottonae spider of Europe spins a web so tiny it's hard to find. It would barely cover a postage stamp.

Water Web

The water spider, as you might guess from its name, doesn't mind the water. But before plunging in, it moves its bell-shaped web through the water to collect air bubbles. When it has enough, the spider heads below in its own little diving bell to look for food.

Spider Olympics

Long-Jump Champ

If you could jump as far as the European jumping spider, you'd be able to leap over five school buses in a row.

Strongest Puller

If the Southern Californian trapdoor spider were your size, it could pull a bus around by its fangs.

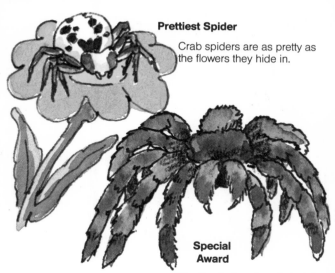

Prettiest Spider

Crab spiders are as pretty as the flowers they hide in.

Special Award

The South American tarantula is big enough to catch birds; it's about the same weight as a juicy plum and it's the hairiest and longest-living spider of them all. There are even a few 30-year-olds creaking around.

Fastest Runner

The female house spider can sprint 5.80 m/19 feet in just ten seconds. That's like you running around the block in ten seconds.

Did You Know?

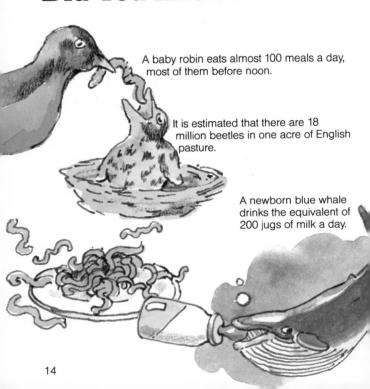

A baby robin eats almost 100 meals a day, most of them before noon.

It is estimated that there are 18 million beetles in one acre of English pasture.

A newborn blue whale drinks the equivalent of 200 jugs of milk a day.

One beaver can cut down 216 trees a year.

About 80,000 bees must fly the equivalent of three times around the world to gather the nectar for a large jar of honey for your toast.

A butterfly flaps its wings about 300 times a minute.

In a Garden by Day

If you go out in a garden on a summer's day, you can meet some very interesting creatures....

Chipmunks can stash up to eight times their weight in seeds in their burrow before hibernating.

A toad is a tidy creature. It gobbles up its cast-off skin while slipping it over its head. Gulp!

Come too close to a garter snake and it will secrete a foul smell from its anal scent glands.

Butterflies never grow. They come out of their cocoons at full size.

In summer, ladybug moms lay 300 small golden eggs—each no bigger than a tiny pinhead.

If you think there are a lot of ants around, you're right. There are a quadrillion (one, followed by 15 zeroes) ants on earth.

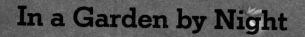

Future Superplants

Cross a banana with a blueberry and what do you get? A banaberry? A bluana? Silly—maybe, but scientists are busy mixing and matching plants in the hopes of developing healthier, bigger ones that grow in unusual places....

If you've ever tried to stack up a pile of watermelons, you'll agree that it's difficult. So here's some good news: a Tokyo designer has successfully grown cube-shaped melons.

Corn plants, growing tall and close together, produce lots of food in a very little space. Sorghum, a kind of grain, grows almost anywhere with little water. So scientists have put the two together to get a plant that can do both things. Hm-m. Humcorn?

What would you call a plant that's half tomato and half potato? A pomato, of course! West German scientists tried to grow a pomato that would produce tomatoes above the ground and potatoes below. They only got a leafy plant, but they're trying again.

Beans, peas and other vegetables in the legume family can produce their own nitrogen—a chemical element that plants need in order to grow. Now scientists are working to fuse cells from legumes with cells from other vegetables. Then they'd all be able to be grown in poor soil.

Kenaf is a plant from Southeast Asia that looks a bit like sugar cane, and can be made into paper for a fraction of the price of making paper from wood.

Scientists are proud of a new kind of potato that can fight off diseases, grow faster than usual and is uniform in shape. So there'll be no more arguments when you share a bag of chips made from such superspuds.

Who's Coming to Dinner?

If you could peek in on an animal dinner party, you'd see not only an unusual menu, but some very amazing eaters....

Where would you look for three overcoats, a pair of pants, a pair of shoes, deer antlers, a cow's hoof, a dozen lobsters and a chicken coop? All these things have been discovered in the belly of one Australian tiger shark.

The blackswallower can gulp fish twice its own size. How? By pushing food down its throat with movable teeth and shifting its heart and gills out of the way.

Instead of delicately picking meat out of a shell with one of its many arms, the sea star has different manners. It opens its mouth and plops out its stomach right on top of its dinner.

Flies don't walk all over your food just for the exercise. They do it to taste what they're about to eat. How? Tiny hairs on their feet act like the tastebuds on your tongue.

Don't stand under a tree if a hungry elephant is nearby. It will knock a tree over if it can't reach leaves from the top.

The plover of Egypt is a plucky bird. It will enter a crocodile's mouth to pick food from the croc's teeth.

When a European mole catches an earthworm, it bites off the top of the worm's head, ties its body into a knot and stores it underground for later.

The red-backed shrike from Europe rarely loses its lunch. That's because after catching insects or other creatures, this bird sticks them onto the spines of the blackthorn bush.

Neat Beaks

Birds' beaks are like a tough combination of lips and jaws with built-in noses. Here are six incredible kinds of bills worth being nosy about.

The anhinga's beak is a super dart. It uses its beak to spear a fish underwater. Then it flips it into the air and swallows it headfirst. This way the fins don't get stuck in the anhinga's throat.

When the skimmer flies low over water, it can pick up fish whenever it wants. It simply drags the long bottom part of its bill below the water. How super!

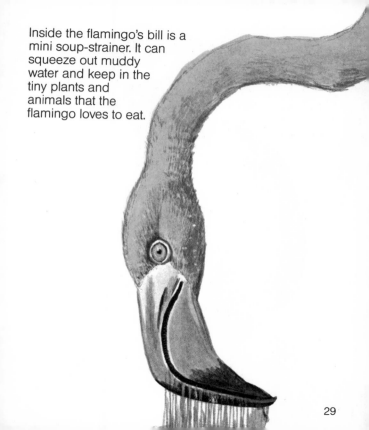

Inside the flamingo's bill is a mini soup-strainer. It can squeeze out muddy water and keep in the tiny plants and animals that the flamingo loves to eat.

The woodpecker's bill helped this bird earn its name. It is long, very strong and chisel-pointed and makes searching for grubs beneath tree bark as easy as tap, tap, tap.

The crossbill's upper and lower bill cross over each other. It looks odd, but it's handy for prying open pine cones and picking out pine seeds.

The tiny hummingbird has a built-in "drinking straw" in the middle of its face. For a sip of nectar, it reaches deep inside a flower and sucks.

Journey through a Dream

by Steven Thornton

MINI MYSTERY

I stepped outside one cool fall morning and saw something red shimmering over the lawn. Slowly, I realized it was a bird—but it seemed to be just hanging there. I could see no wings—how did it stay up?

Then the answer came to me. My hovering friend was a male ruby-throated hummingbird, a tiny nectar-feeder with a drinking straw for a bill and a name taken from the iridescent red feathers on his throat.

He had wings all right, but they were beating so fast—50 times a second—that they were as invisible as the whirling blades of a helicopter. And like a helicopter, he could fly in one spot, looking as if he was suspended from the sky. Finally he did move—backwards. As I blinked in astonishment, he vanished.

At that instant, my ruby-throat had darted off to begin an amazing journey full of danger and excitement. I couldn't join his migration, but I could follow him in my mind. This is our adventure....

When I first glanced down, I nearly fell out of the sky! The cars on the road below were like tiny toys. We followed them for a time, flying the same speed, then we banked over a forest and came to a field where flowers beckoned far below. My friend circled downwards and I followed. Hovering in front of a flower, he inserted his long, thin bill into the blossom and sucked out the sweet nectar, then, moving sideways, flew to another flower. Darting from blossom to blossom, he fed himself then napped until day became night, and we were off again. For an entire month our routine was the same: flying in darkness, then stopping for a day or two while my friend nourished himself to build strength for the next portion of the long journey. Why would this bird make such a long trip? I knew that my hummer's thin coat of feathers couldn't protect him against the cold northern winter, and of course, he couldn't find nectar in frozen blossoms. It made sense to go south to Central America where flowers bloom all year—but why not just stay there? Maybe some species of hummingbirds had discovered that by moving north in the spring, they would have less competition for flowers.

I was pondering this while flying along one night when a roaring, rushing sound shook my thoughts right out of my head. The stars faded and a sudden lightning bolt split the darkness: a hurricane was upon us. My hummer and I tumbled across the sky, bouncing on currents of air and lashed by bullets of water.

When at last I came to I wondered: was I lost? Was I dead? Where was my friend? What seemed like a lifetime later, I found myself surrounded by something white—fog. Through the vapor, I saw a patch of color and sped toward it. And drawing his morning nectar from a flower as if nothing had happened, was my friend.

Once the fog lifted, the sun burned in a blue sky. We were far south now, I thought. Florida, or Louisiana, by the Gulf—and we'd come all the way from Alberta. A warm breeze washed over us. Several days later, as the sun was setting, my friend darted into the air. He appeared to be studying the sun's position—nervously trying to figure out where we were, and where we must go.

As soon as it got dark we took off, flying low, and as the moon rose I saw water below: the Gulf of Mexico. We flew on for hours. Then as the morning light spread out before us, I saw the coast of Central America. There lay flowers enough for my friend's winter, tropical warmth and thousands more hummers. We had come a long, dangerous way, and now his winter home was in sight.

Suddenly feeling sad, I looked at my ruby-throat. It was time for us to part. In a few minutes he would begin dropping down through the sky toward the fields and the others of his kind would be waiting below. I would leave, and never see him again.

Or would I?

I closed my eyes, and when I opened them again I was back in my front yard, breathing in the brisk autumn air. I was alone now; my friend was far away. But he would be back, I knew it. I went inside and began counting the days until spring.

High Flying Olympics

Most Versatile

The hummingbird can fly backward—the only bird in the world that can.

Fastest Flapper

The horned sungem of tropical South America can flap its wings 90 times a second. How many times can you wave your arms in that time?

Biggest Wingspan

A big wave to the wandering albatross. It has a whopping wingspan of just over 3 m/10 feet.

Highest Flyer

In 1967 an airline pilot was surprised to see 30 whooper swans outside his window. He—and they—were flying 8,230 m/27,000 feet high.

Long-Distance Champ

Hats off to the arctic tern. No sooner does it raise its family in nesting grounds close to the North Pole than it turns around and flies south to its winter feeding grounds—all the way to the South Pole.

The Fastest

The white-throated spine-tailed swift soars to first place as fastest bird at 169 km hp/105 miles per hour.

Most Feathers

At last count the whistling swan was found to have 23,216 feathers...

Weather Wisdom

Insects buzz close to the ground before a rainstorm because their wings are heavier than normal with dampness. And when insects fly low, so do hungry swallows.

The water-swollen dust particles that fill the air just before a storm block out blue and yellow light rays from the sun. Only the very long, deep red rays get through and in the morning that makes "sailors take warning."

The crackling noise you hear on the radio or TV during a storm is caused by lightning zapping from the clouds to the ground. If you can't see lightning and you still hear static, the lightning is probably inside the clouds.

A bright halo around the sun or moon is caused by light shining on ice crystals in very high clouds. When clouds have been pushed high it often means a heavy rain—or a snowstorm—is on the way.

Ever hear the saying "it smells like rain?" That's the smell of rotting vegetation and perfume from flowers. Just before a rainstorm, the air around us presses down less heavily on the ground than normal. So smells have a chance to rise higher.

It's Raining Cats and Dogs

Why do people use the expression "raining cats and dogs"? No one knows, but we do know some amazing facts about rain.

An average-size thundercloud holds about six trillion raindrops.

There are about 2,000 thunderstorms raging around the world every second of the day.

You can measure rainfall by putting a glass outside in a storm. See how close you are to this record rainfall: almost 3.8 cm/1.5 inches fell in one minute on November 26, 1970, in Guadeloupe.

If you love rain, Mt. Waialeale, Kauai, Hawaii, is the place for you. It rains 350 days each year on average.

People living in Cilaos on La Réunion Island in the Indian Ocean found themselves waist deep in water one day in 1952. That is because 187 cm/73.62 inches of rain fell on them in 24 hours.

Most rainbows last about 30 minutes. Some folks in North Wales almost had time to find the pot of gold one day in 1979—their rainbow lasted three hours!

Cats Head to Toe

Most cats have 20 muscles or so in each ear so they can swivel them both in almost any direction.

Cats' eyes face forward so that what each eye sees overlaps, giving them "stereo" vision just like yours, only better.

Cats, like many animals, have a special "third" see-through eyelid next to the eye. Closing this keeps the eye safe from dust and properly lubricated.

A cat's stiff, wiry whiskers are sensitive feelers that tell it how much room there is around its head.

A cat smells first with its nose, but to identify what it's smelling it opens its mouth and lifts its head to let the scent waft into a small pouch—called the Jacobson's organ—in the roof of its mouth.

The leatherlike nose pad you see on a cat is a protective covering for its smelling apparatus.

The comblike spikes—papillae—on a cat's tongue are used as a fur brush.

Cats cannot move their jaws sideways to chew so they must swallow their food in small chunks.

Even the tiniest of cats has more bones than you do: 230 compared to your 206. Twenty of them are in the tail.

A cat's fur, consisting of large guard hairs and shorter underfur, protects its body from bites, scratches, heat and cold. Each guard hair is attached to a muscle that makes it stand on end whenever the cat is angry or alarmed.

The muscles in a cat's legs are very strong, making it a good jumper, sprinter and climber.

When a cat isn't using its needlelike claws for climbing, digging or fighting, it can pull them up out of the way.

Cat Ways

Visit the natural kingdom to watch the behavior of wild cats and you'll understand much more about why your cat does the things it does.

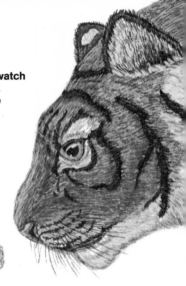

A leopard is very good at hitching its way up a tree using its strong claws as hooks. But coming down is a different matter. Even your house cat can't get a good grip on the way down. No wonder it will usually jump or wait to be rescued.

Most cats—wild or domestic—stake out their territory by scratching, urinating or rubbing against things to leave a scent. The size of a cat's territory depends on the food supply—more food, less space.

Wild cats are careful about washing after meals because leftover blood may attract flies and enemies. Domestic cats still have the habit of washing by moistening their favorite front paw and using it as a facecloth.

It's easy to tell if tracks are made by a cat: you can connect them with a straight line. That's because all cats move the front and back legs on one side, then the front and back legs on the other. See?

Cat Chat

Cats can tell you a lot if you know their language....

A loud meow generally means "Listen to me, I want something."

A gently waving tail or tail held upright means a cat is pleased.

No one is quite sure how a cat purrs, but this noise is mostly a sign of contentment.

Watch out for a swishing tail. This is a first sign of a cat's anger. A growl will come next.

Ears that are bent forward mean a cat is curious; held out to the sides, nervous; flattened back, angry.

That loud caterwaul you hear at night is a mating call.

Only a secure, happy cat will sleep in a "belly-up" position.

When cats feel friendly, they touch noses and whiskers.

Wild Cat Olympics

Smallest

The rusty spotted cat of Sri Lanka is not much bigger than the birds it catches.

Most Agile

The African caracal has such lightning-fast reflexes that it can grab birds out of the air.

Biggest

Even elephants fear the Siberian Tiger. It's the largest and most powerful cat of all.

Best Swimmer

No wonder the Asian fishing cat is an excellent swimmer—it has webbed feet.

Koska's Trek

by Steven Thornton

This remarkable story of a six-year-old house cat began on a pleasant but uneventful summer vacation. Koshka and her people were driving through the mountains of southeast British Columbia when she jumped out of the car. A few minutes later she found herself alone on an empty highway, her home far away across the rough-and-tumble peaks. If she could tell us what happened next, this is what she might have said.

Almost immediately I knew I was in trouble. Those high mountains, all covered in dark and tangled forests, were no place for a house cat.

I plodded over to the edge of the road and jumped down onto a grassy hillside. Everything was so bewildering. Where was my family? Why did they leave me? How would I get home? After a while, however, the warm sunshine made me feel better, so I ran fast, my tail high.

Following the twisting banks of a stream, I came to a forested slope. On I tramped, under trees dressed in long green needles, and upward through leafy ferns crowding the gloomy forest floor. Finally I found myself high on a cliff overlooking a valley with purple mountains that towered over the far side. Below was a river, flashing like a silver ribbon, and a little village. Somehow, I knew that my way home lay through the valley and over the mountains.

First, though, I had to reach the narrow ledge angling down across the cliff's face. I stretched forward and almost touched it—and then one more try and I was safely there. The cold wind brushed my fur and there was no sound as I crept cautiously on. The rock ahead looked secure as I stepped onto it….

I wailed, clawing at the air as I plunged upside down through space. Snapping my head around, I got turned over, but falling seemed to take forever—as if in a dream. I banged against another ledge and somersaulted away.

Finally, I crashed into water that felt hard as rock. It knocked the breath out of me, but, sputtering and coughing, I managed to swim. The water was so cold it burned, and by the time I reached the bank I was numb.

I dragged myself out and lay down. I hurt, but nothing was broken, so when I felt better, I limped into a sunny meadow to rest a bit more. How happy I felt to be on flat ground again.

Then I saw something that sent a tingle through me. High up on a branch of a scraggly tree, with its back to me, sat a chipmunk. I may be a pampered house pet, but when I saw that chipmunk, I changed. Now I was a quick, silent hunter, gliding forward with my ears flat on my head. At the base of the tree I paused for a last, calculating look at my prey.

With a burst of power I leapt into the tree and clawed my way up. As I reached its branch, the chipmunk spotted me and jumped straight up in alarm. But when it came back down I was there, and in a single bite I ended its struggles.

Standing on that branch with my prey locked between my jaws, I took a moment to relax. Now I knew that I was going to make it home. I had survived a fall from an impossible height into agonizingly cold water, and found food. I had become a wild cat, fit for a wild country. In the village below, I would be tame again, seeking food from the people there. But when I left the village, I knew I could throw off my tameness. The mountains beyond the valley no longer seemed so forbidding.

I descended the tree and ate my meal slowly. When I was done, I moved onward.

It took six months, but Koshka did get home. Undoubtedly, there were many other dangers in her journey, but whenever the odds seemed too great, she somehow found the courage to take them on. By winter Koshka rejoined her family and became a house pet once again.

What's in a Name?

Remember the dwarfs in "Snow White"—Doc, Bashful, Happy, Sleepy, Sneezy, Dopey and Grumpy? Their names described how they behaved. Many animals also have names that tell something about them. Amuse yourself by guessing a few.

Answers on page 95

I got my name because I shoot insects with drops of water. I can knock a beetle into the water from halfway across a small pond. I rarely miss my target.

I'm an _____ fish.

I burrow into the ground with scooped front legs and sharp claws shaped like the paws of a mole.

I'm a _____ cricket.

When I try to attract a female into my burrow I look as if I'm bowing a fiddle. But I'm really moving my large claw slowly back and forth.

I'm a _____ crab.

You'd be excused for not noticing me when I'm on my back. That's because I look like a piece of wood. But when I turn over—with a click!—I can skip into the air and land on my feet.

I'm a _____ beetle.

I'm equipped with a built-in fishing rod and bait. A blob of muscle dangles from a long, thin bone on my head. When it wiggles, small fish come to investigate, only to be snapped into my wide mouth.

I'm an _____ fish.

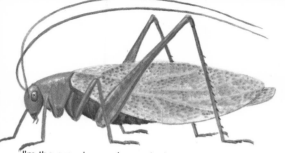

I'm the grasshopper's cousin. I call my lady-friend at night by rubbing my wings together to make a sound like "katy-did, katy-did."

I'm a _____.

I live in Australia and am named after the wide, pleated frill around my neck and shoulders. When I'm in danger, I fan this frill out. It's all a bluff, however. I'm really quite harmless.

I'm the _____ lizard.

Did You Know?

A leaky faucet drips enough water in one day to fill six bathtubs. If you can put a brick in your toilet tank (very carefully), you can save one or more large juice cans of water with each flush.

About 40 fireflies in a dark room will make enough light to read by.

When you save a tankful of gas in your car, you're saving enough petrochemicals to manufacture 14 bicycle tires.

Energy saving is on most people's minds. But in some places, like West Germany, it's on a few people's roofs. They're planting grass on the top of their houses. The soil underneath stores the sun's heat during the day, and at night a special pump circulates it around the house. Roof-top sheep cut the grass.

Zany Olympics

What would happen if humans competed with mammals, insects, birds and fish for some sports championships? The marathon would probably be the major event, with prizes being given to those who can keep moving nonstop for the longest time. The results would be as follows....

Bronze Medal

With its small, weak legs, the common swift is helpless on the ground and easy prey for enemies. It's not surprising, therefore, that it spends up to nine months a year in the air, landing only to lay its eggs and raise its young.

Silver Medal

After a young sooty tern leaves its nest it flies nonstop for up to four years before it returns again to its breeding grounds in the tropical Atlantic, Pacific or Indian oceans. Experts aren't sure why this happens, but they agree that the tern is some flier.

Gold Medal

Three fish—the great white shark, the blue fin tuna and the swordfish—tie for first place. They keep swimming all their lives, for if they stop, they die. They can live up to 30 years. That's some swim.

THE DEEP DIVE

The deep dive would be an interesting event, with competitors, big and small, being judged on the depth reached. Here are the winners...

Bronze Medal

The sea lion dives 600 m/1,968 feet.

Silver Medal

The sperm whale often dives to 1,134 m/3,719.5 feet, but it may dive to 3 km/1.9 miles looking for delicious squid.

Gold Medal

In 1960 two men in a small submarine went down 10.8 km/6.7 miles under the sea.

IT SURE IS DARK DOWN HERE.

GOLD MEDAL

THE HIGH JUMP

Prizes for high jumping go to competitors that jump the highest in proportion to their own size.

Gold Medal
The tiny flea can jump 130 times its own height. A human would have to leap over a 65-story building to jump as high.

Silver Medal
The flying fish can make a remarkable leap out-of-the-water to a height of 20 times its own length.

Bronze Medal
The kangaroo can jump about 1 m/39 inches more than its height.

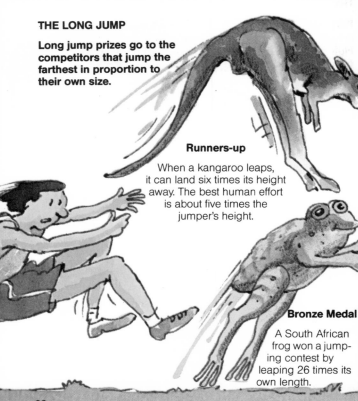

THE LONG JUMP

Long jump prizes go to the competitors that jump the farthest in proportion to their own size.

Runners-up

When a kangaroo leaps, it can land six times its height away. The best human effort is about five times the jumper's height.

Bronze Medal

A South African frog won a jumping contest by leaping 26 times its own length.

Gold Medal

The flying fish has winglike fins so it's hard to tell when it's jumping and when it's flying. Nonetheless, it can travel the length of three football fields.

Silver Medal

The amazing flea can jump 200 times its own length. If an adult man could do this he would end up about five city blocks away.

Foot Notes

Humans have two feet. Most creatures have more—for instance, one southern European centipede has 177. But your feet are among the world's most interesting…

During a lifetime, an average person walks the equivalent of five times around the earth at the equator.

Not only fingerprints reveal a person's true identity. Your toe- and foot- prints are unique too. Off with your socks!

If your feet are cold, put a hat on your head. By keeping your head warm, you keep your brain warm so your body will send blood to such far places as your feet.

Stamp your feet for Kumar Anadan of Colombo, Sri Lanka. He wanted to set a record so he stood on just one foot for 33 hours in May, 1980.

The Bedford North Scout Group of England may not want to see another foot again. In 1977 they broke the world record when they shined 6,334 pairs of shoes in eight hours.

Johann Hurlinger of Austria loved to walk, but not on his feet. To get from Vienna to Paris in 1900 he started walking on his hands— and made it!

Feat Notes

Happy birthday to you and you and you. You share your birthdate with at least nine million other people in the world.

Mr. and Mrs. Ralph Cummings of Clintwood, Vermont, are unusual parents. Their five children were all born on February 20, but in different years. The chances of this happening in a family is one in almost 18 billion!

All hamsters in North America are descended from just three golden hamsters raised and bred by an Israeli scientist in the 1930s.

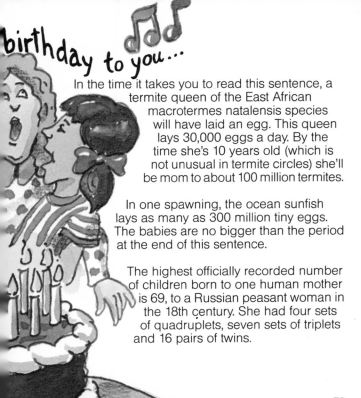

birthday to you...

In the time it takes you to read this sentence, a termite queen of the East African macrotermes natalensis species will have laid an egg. This queen lays 30,000 eggs a day. By the time she's 10 years old (which is not unusual in termite circles) she'll be mom to about 100 million termites.

In one spawning, the ocean sunfish lays as many as 300 million tiny eggs. The babies are no bigger than the period at the end of this sentence.

The highest officially recorded number of children born to one human mother is 69, to a Russian peasant woman in the 18th century. She had four sets of quadruplets, seven sets of triplets and 16 pairs of twins.

Superparents

In the sky...in the water...on land...here are some of the most amazing parents of the animal world.

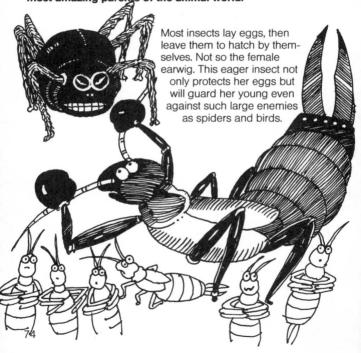

Most insects lay eggs, then leave them to hatch by themselves. Not so the female earwig. This eager insect not only protects her eggs but will guard her young even against such large enemies as spiders and birds.

After the male sea horse fertilizes his mate's several hundred eggs, his job has just begun. Then he must scoop the eggs up into his pouch and keep them safe until they hatch.

So as not to eat their babies by mistake, mother sharks will not eat anything in the area where they have just given birth.

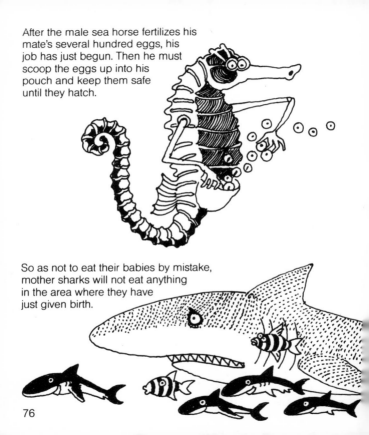

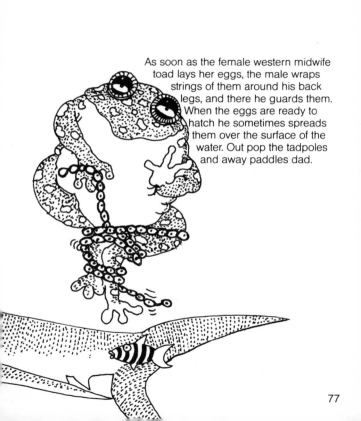

As soon as the female western midwife toad lays her eggs, the male wraps strings of them around his back legs, and there he guards them. When the eggs are ready to hatch he sometimes spreads them over the surface of the water. Out pop the tadpoles and away paddles dad.

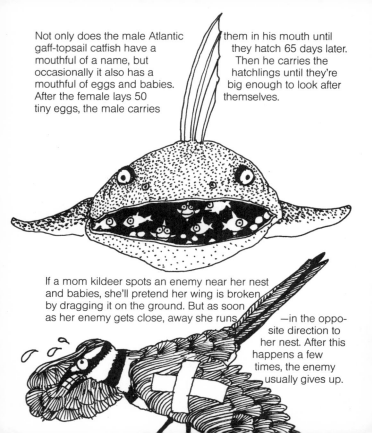

Not only does the male Atlantic gaff-topsail catfish have a mouthful of a name, but occasionally it also has a mouthful of eggs and babies. After the female lays 50 tiny eggs, the male carries them in his mouth until they hatch 65 days later. Then he carries the hatchlings until they're big enough to look after themselves.

If a mom kildeer spots an enemy near her nest and babies, she'll pretend her wing is broken by dragging it on the ground. But as soon as her enemy gets close, away she runs —in the opposite direction to her nest. After this happens a few times, the enemy usually gives up.

Once crocodile babies crawl out of their nest, they crawl into their mother's mouth. She carefully carries them to a safe nursery spot where she'll look after them for up to six months.

Did You Know?

If you're ever sitting on an African river bank and a goliath frog hops by, move over. This monster frog is over twice the size of a robin.

The granddaddy of insects, the goliath beetle, is as big as your fist. It hangs around in trees in Africa and eats only plants. Thank goodness.

The giant squid has the world's biggest eyes—they're about the size of dinner plates. The better to see you with.

More than 20 children holding hands could barely manage to circle the giant sequoia tree of California. It's the biggest living thing in the world.

The ostrich is such a big bird that it would bump its head if it came through your front door.

The biggest animal in the world, the blue whale of the Atlantic and Pacific oceans, has a heart the size of a subcompact car.

The bald eagle has been known to build a nest as deep as a swimming pool and as wide as three bathtubs.

Real Monsters

The biggest lizard in the world looks like an honest-to-goodness dragon. It's the size of a small horse and can chase down wild pigs and swallow them whole. No wonder when early visitors to the South Pacific island of Komodo first described this monster, people would hardly believe them. Some people think that there are other monsters in the world that are just as real but harder to find than the Komodo. Turn the page and meet some land monsters people think are real….

North America's Monster

Some people call the gigantic, hairy, two-legged creature reported as seen in the west a sasquatch. Others call it Bigfoot. Either way, its footprint is supposed to be twice as big as one made by an average man.

Experts think an adult sasquatch is twice as tall and three times as heavy as an average adult person.

Apparently you know when a sasquatch is coming by its very strong smell.

Seen One Lately?

If the sasquatch exists at all, it certainly seems to be a very shy creature that tries to avoid people. But sometimes it hasn't been too successful...

Around 1900 a Chehalis Indian said he was chased home by a giant that was wading in a small river. The tribe has never forgotten either the giant or the Indian who made the claim.

Workers building the Canadian Pacific Railway in 1884 captured a "gorilla-like" creature by dropping rocks on it from above. They called it Jacko and planned to send it to England. It never seemed to arrive. Did Jacko escape or was he sold to a circus? Or did Jacko ever exist at all?

Albert Ostman, a prospector, reported being carried off by a strange creature in 1924. After spending several days as a prisoner with the creature's family of four, he escaped. But he did not tell his story for 33 years for fear of being mocked.

In 1977 a bus full of people saw what they believed was a sasquatch standing on the highway. Later, however, three men admitted the whole thing was a hoax and that one of them had dressed up in a monkey suit.

Californian Roger Patterson was horseback riding in 1967 when a dark animal made his horse rear. After being thrown to the ground, he grabbed his camera and started filming just as the creature fled into the bush. Afterward, Patterson discovered he had a movie of what looked like the sasquatch.

Look for more sasquatch/Bigfoot sightings in your newspaper to add to your collection. You never know when one might turn up again.

China's Wild Man

If you can't decide whether the sasquatch—or Bigfoot—exists, maybe you'll have better luck with the strange Wild Man of China, or Yeh ren, as it's locally known. Chinese scientists have launched several expeditions into the forests near the Yangtze River to capture it alive.

Though they haven't been successful, they have found some interesting bits of evidence that suggest Yeh ren exists....

Scientists have found a cave where they say an apelike creature lives.

A peasant woman reported seeing a strange, hairy animal rubbing its back against a tree. Scientists later found hairs on this tree. When they examined them, they resembled hair of higher primates, including man.

Scientists have talked to people who say the Yeh ren walks on its back legs and is about the size of a tall person.

Its face looks like a human's but it has bigger ears, a wide mouth and large jaw. It is covered with reddish-brown hair.

Like the sasquatch, the Yeh ren has feet about twice as big as those of an average adult human.

Yeh ren has been known to clap its hands, smile and imitate the cries of other animals, including the chirping of sparrows, the barking of dogs, the crying of babies and the noises of leopards.

Some scientists think the Yeh ren is a cross between a gorilla and a human being; others suggest it is a rare Chinese golden monkey. Until they capture one nobody will know for sure.

Yeti of the Himalayas

You can't call yourself a monster lover without knowing something about the yeti, or abominable snowman, as some call him. Many people claim they've seen this two-

legged giant wandering in the Himalayan mountains, but experts have only seen footprints. They say they're *three* times bigger than a human's foot.

The Chinese wrote about the yeti as early as 200 B.C. One of the first reported yeti sightings by a Western explorer was made by William Knight in 1903. He describes the yeti as a tall, pale yellow creature without much hair on its face. It was carrying what looked like a bow.

Kangaroo
  high jumper, 67
  long jumper, 68
    Kenaf (plant), 22
Kildeer (bird), 78
Komodo dragon, 82-83
"Koshka's trek," 54-57

Ladybugs, 17
Largest animal, 81
Legumes, 22
Leopard, 48
Lightning, 38
Lizard, 82-83

Melons, 20
Mercaptan (skunk odour), 18
Mice, 19
Moles, 26
Monsters, 82-91
  Komodo dragon, 82-83
  Sasquatch, 84-87
  Yeh ren, 88-89
  Yeti, 90-91

Names, animals, 58-61
Nest, largest, 81

Ocean sunfish, 73
Ogre-faced spider, 9
Orbweaver spider, 10
Ostrich, 81
Owls, 18

Parents, 74-79
Plants, 20-23
Plover, Egyptian (bird), 26
"Pomato," 21
Potato, 23
Purring, 50

Raccoons, 19
Rain
  most rain in year, 41
  record rainfalls, 40, 41
  signs of rain, 38, 39
  thunderstorms, 40
Rainbow, longest, 41
Robin, feeding habits, 14

Sasquatch, 84-87
Sea horse, 76
Sea lion, 66
Sea star, 24
Sequoia tree, 81
Shark
  contents of belly, 24
  long swimmer, 65
  as mother, 76
Shoe shining, 71
Shrike, red-backed (bird), 27
Siberian Tiger, 53
Skimmer (bird), 28
Skunk, 18
Smell
  garter snake, 17
  rain, 39
  skunk, 18

whiskers, 42
wild, 46-49
Children, 73
Chipmunks, 16
Conservation
   energy, 63
   gas, 63
   water, 62
Crab spider, 13
Crocodiles, 79
Crossbill (bird), 31

Dive, deepest, 66

Earwig, 74
Egg-laying, 73
Elephant, 26
Energy conservation, 63
Eyes, 80
Faucet, 62
Feet, 70-71
Fireflies, 19, 62
Fish
   as parents, 76, 78
   eating habits, 24
   egg-laying record, 73
   high jump, 67
   long jump, 69
   longest swim, 65
Flamingo, 29
Flea
   high jumper, 67
   long jumper, 69
Flies, 26

Flying fish
   high jumper, 67
   long jumper, 69
Frog, 68

Garter snake, 17
Gas, 62
Giant squid, 80
Glyphesis cottonae spider, 11
Goliath beetle, 80
Goliath frog, 80
Grass on roofs, 63
Great white shark, 65

Halo around the sun or moon, 38
Hamsters, 72
Hornbill (bird), 75
Horned sungem (bird), 36
House spider, 13
Humans
   deepest divers, 66
   foot notes, 70-71
   most children born to one
     mother, 73
Hummingbird
   backward flying, 32, 36
   bill, 31
   migration, 32-35
   wing speed, 32

Jacobson's organ (cats), 43
Jumping, 67, 68-69
Jumping spider, 12

Index

Almas (wild men), 91
Anhinga (bird), 28
Ants, 17
Arctic tern, 37
Asian fishing cat, 53
Atlantic gaff-topsail catfish, 78

Bald eagle, 81
Beaks, 28-31
Beavers, 15
Bees, 15
Beetles, 14
Bigfoot (monster), 84-87
Birds
  backward-flying, 36
  beaks, 28-31
  biggest wingspan, 36
  crocodiles' toothpicks, 26
  fastest, 37
  fastest wing speed, 36
  food storage, 27
  highest-flying, 37
  hunger, 14
  keen hearing, 18
  largest, 81
  largest nest, 81
  longest flights, 37, 64, 65
  most feathers, 37
  parents, 75, 78
Birthdays, 72
Blackswallower (fish), 24

Blue fin tuna, 65
Blue whale
  drinking capacity, 14
  largest animal, 81
Bolas spider, 8
Bones, 44
Butterflies
  never grow, 17
  wing speed, 15

Caracal (African wild cat), 52
Cats, 42-57
  best swimmer, 53
  biggest, 53
  bones, 44
  claws, 45
  climbing, 48
  ears, 42
  eating, 44
  eyelids, 42
  eyes, 42
  fur, 45
  Jacobson's organ, 43
  "Koshka's trek," 54-57
  "language," 50-51
  most agile, 52
  muscles, 45
  nose pad, 43
  scent, 43
  smallest, 52
  staking out territory, 48
  tongue, 44
  tracks, 49
  washing, 49

Some experts think the yeti are really langur monkeys occasionally found high in the Himalayas.

Yeti apparently also live on the Russian side of the Himalayas, where they're called almas.

The big question about all the strange creatures of the world is, do people see them because they're there or because they want to see them? Nobody knows that either.

Sooty tern, 65
Sperm whale, 66
Spiders, 6-13
   fastest runner, 13
   feet, 10
   longest-living, 13
   long jump champ, 12
   prettiest, 13
   strongest, 12
   webs, 10-11
Squid, 80
Superplants, 20-23
Swift
   common, 64
   white-throated
   spine-tailed, 37
Swordfish, 65

Tarantula, 13
Termites, 73
Thunderstorms, 40
Tiger shark, 24
Toads
   as parent, 77
   cast-off skin, 16
Trapdoor spider, 12
Trees, 81

Walking on hands, 71
Wandering albatross, 36
Water, 62

Water spider, 11
Weather, 38-41
   signs of rain, 38, 39
Webs, spider's, 10-11
Western midwife toad, 77
Whale
   drinking habits, 14
   largest animal, 81
Whistling swan, 37
White-throated spine-tailed
   swift, 37
Whooper swans, 37
Wild cats, 46-49
Wild men, 84-91
Wolf spider, 6-7
Woodpeckers, 30
Yeh ren (wild men), 88-89
Yellow spitting spider, 8
Yeti (wild men), 90-91

Answers

What's in a Name.?
Pages 58-61

1. archer
2. mole
3. fiddler
4. click
5. angler
6. katydid
7. frilled

Illustrations:
pp
2-3, 4, Joe Weissmann
6-7, 8-9, Julian Mulock
10-11, 12-13, 14-15,
   Joe Weissmann
16-17, 18-19, Debie Perna
20-21, Tina Holdcroft
22-23, Joe Weissmann
24-25, Ron Taylor/
   Bruce Coleman Inc.
26-27, Lynda Cooper
28-29, 30-31, Anker Odum
32-33, Elaine Macpherson
36-37, Joe Weissmann
38-39, 40-41, Tina Holdcroft
42-43, 44-45, Lynda Cooper
46-47, David Grainger
48, Alan Daniel
49, 50-51, Lynda Cooper
52-53, Joe Weissmann
56, Olena Kassian
58-59, Julian Mulock
60-61, Elaine Macpherson,
   Julian Mulock
62-63, 64-65, 66-67, 68-69,
   Joe Weissmann
70-71, Tina Holdcroft
72-73, Joe Weissmann
74-75, 76-77, 78-79, 80-81,
   Tina Holdcroft
82-83, Olena Kassian
84-85, Clive Dobson
87, Patterson, Gimlin
88-89, Tina Holdcroft
90-91, Clive Dobson

Credits
Special thanks to: Dr. A Baker,
D. Barr, Dr. C. Dundale,
Dr. A. Emery, Dr. J. Grayson,
R. James, F. Larsen,
K. MacKeever, R. McCollough,
Dr. R. Schemanaur and all
the people at the Royal Ontario
Museum, Ontario Science
Centre and Metropolitan
Toronto Zoo.

Edited by Laima Dingwall
& Annabel Slaight
Cover designed by
Wycliffe Smith
Cover illustration by
Vesna Krystanovich